HARCOURT

Math

Performance Assessment

Grade 2

Harcourt

Orlando Austin Chicago New York Toronto London San Diego

Visit *The Learning Site!*
www.harcourtschool.com

Printed in the United States of America

ISBN 0-15-336856-X

3 4 5 6 7 8 9 10 022 10 09 08 07 06 05 04

CONTENTS

Performance Assessment Program

Unique Features of *Harcourt Math* Performance Assessment

To create assessments that actually evaluate what is taught, it is necessary to target specific math concepts, skills, and strategies at each grade level. In planning the assessment program for *Harcourt Math,* a review was made of performance assessments cited in professional literature and also of those used in state testing programs. Comparisons were made among available models, and desirable features were identified. Holistic scoring was chosen as the primary method of scoring. The *Harcourt Math* performance assessments offer the following features:

- **They model good instruction.** The assessments are like mini lessons.
- **They are diagnostic.** By reviewing students' notes, teachers gain valuable insight into the thinking strategies that students are using.
- **They encourage the thinking process.** The assessments guide students through the process of organizing their thoughts and revising their strategies as they solve problems.
- **They are flexible.** No strict time limits are imposed, and students are encouraged to proceed at their own pace.
- **They use authentic instruction.** Each task is based on realistic problem solving situations.
- **They are scored holistically.** Responses are scored by a multidimensional rubric to provide a comprehensive view of each student's performance.

Development of the Performance Assessment Program

Each assessment was field-tested with students before it was selected for inclusion in the program. After the assessments were selected, the pool of student papers for each assessment was reviewed, and model papers were selected to illustrate the various levels of performance. Annotations were then written for each model paper, explaining why the particular level was assigned.

The development process provided an opportunity to drop or correct those assessments that were not working as expected.

Administering the Performance Assessments

- **Be encouraging.**
  Your role in administering the assessments should be that of a coach—motivating, guiding, and encouraging students to produce their best work.

- **Be clear.**
  The directions for the assessments are not standardized. If necessary, you may rephrase them for students.

- **Be supportive.**
  You may assist students who need help. The amount of assistance needed will vary depending on the needs and abilities of your students.

- **Be fair.**
  Allow students adequate time to do their best work. They should not feel that they could have done better if they had been given more time.

- **Be flexible.**
  All students need not proceed through the assessments at the same rate and in the same manner.

- **Be involving.**
  Whenever possible, involve students in the evaluation process.

Providing for Students with Special Needs

Many school districts are facing the challenge of adapting instruction and assessment to make them appropriate for their learners with special needs. Because the performance assessments are not standardized, the procedure for administering them can be adjusted to meet the needs of these learners. Teachers can help students who have difficulty responding by

- pairing a less proficient learner with a more proficient learner.
- encouraging students to discuss their ideas with a partner.
- providing an audiotape of the performance assessment and having students read along with the narration.
- permitting students to tape-record their responses in lieu of writing them.
- allowing students to do their initial planning, computing, designing, and drafting on the computer.
- giving students extra time to do their planning.
- providing assistance upon request.

Keep in mind, however, that the more the performance assessments are modified, the less reliable they may be as measures of students' mathematical ability.

Scoring Rubrics for Mathematics

In scoring a student's task, the teacher should ask two questions: *How well did the student use the conventions of mathematics to arrive at a solution?* and *How well did the student communicate the solution?* The scoring system used for the performance assessments is designed to be compatible with those used by many state assessment programs. Using a 4-point scale, the teacher classifies the student's performance as "excellent," "adequate," "limited," or "little or no achievement."

4-Point Scale			
Excellent Achievement	Adequate Achievement	Limited Achievement	Little or No Achievement
3	**2**	**1**	**0**

A Level 3 paper shows evidence of extensive understanding of content and provides an exceptionally clear and effective solution. A Level 2 paper shows an acceptable understanding of content and provides a solution that shows reasonable insight. A Level 1 paper shows partial understanding and is clear in some parts, but not in others. A Level 0 paper demonstrates poor understanding of content and provides a solution that is unclear.

Sharing Results with Students and Families

The performance assessment can provide valuable insights into students' mathematical abilities by revealing how all students performed on a common task. However, it is important that their performance on the assessment be interpreted in light of other samples that have been collected such as daily papers, student portfolios, and other types of tests, as well as teacher observation.

For Students

Discuss the rubric with students and explain how it will be used. You may even want to score some anonymous papers as a group or have students score each other's papers and discuss the criteria as they apply to those papers. Make photocopies of the rubrics to use for individual reports. Discuss the reports in conferences with students, pointing out their strengths as well as areas in which they could still improve.

For Families

Results of performance assessments may also be shared with families, who will appreciate seeing what their children can do. Show family members the performance assessment so that they understand the task that the students were asked to perform. Show their child's responses and discuss the strengths and weaknesses of the responses. Explain the scoring rubric and how the responses were evaluated. Show model papers that illustrate the range in student performance to help them put their child's paper in perspective.

Using Results to Assign Grades

No single test, whether a standardized achievement test, a performance assessment, or an open-ended test, can fully measure a student's mathematical ability. For this reason, it is important to use multiple measures of assessment. Therefore, a score on performance assessment should not be used as the sole determiner of a report-card grade or semester grade. The performance assessment could represent one of several factors used to determine a student's grade. Assessments could be combined with the results of a selection of tests, daily grades, class participation, self-reflections, and various samples collected in a portfolio. The following table shows how holistic scores can be converted into numerical or letter grades.

Holistic Score	Letter Grade	Numerical Grade
3	A	90-100
2	B	80-89
1	C	70-79
0	D–F	69 or below

Developing Your Own Rubric

A well-written rubric can help teachers score students' work more accurately and fairly. It also gives students a better idea of what qualities their work should exhibit. Using performance assessment to make connections between teaching and learning requires both conceptual and reflective involvement. Determining criteria may be the most difficult aspect of the process of developing assessment tools with which to evaluate students' performance. Particularly challenging is the task of finding the right language to describe the qualities of student performance that distinguish mediocre and excellent work. Teachers should begin the process of developing rubrics by

- gathering sample rubrics as models to be adapted as needed.
- selecting samples of students' work that represent a range of quality.
- determining the qualities of work that distinguish good examples from poor examples.
- using those qualities to write descriptors for the desired characteristics.
- continually revising the criteria until the rubric score reflects the quality of work indicated.

To develop your own rubric, you may wish to use a format similar to the models on the next two pages.

Your Own Scoring Rubric

Response Level	Performance Indicators
3	Generally accurate, complete, and clear
2	Partially accurate, complete, and clear
1	Minimally accurate, complete, and clear
0	Not accurate, complete, and clear

Harcourt Math Scoring Rubric

Response Level	Performance Indicators
3	**Generally accurate, complete, and clear** _____ All or most parts of the task are successfully completed; the intents of all parts of the task are addressed with appropriate strategies and procedures. _____ There is evidence that the student has a clear understanding of key concepts and procedures. _____ Student work and explanations are clear. _____ Additional illustrations or information, if present, enhance communication. _____ Answers for all parts are correct or reasonable.
2	**Partially accurate, complete, and clear** _____ Some parts of the task are successfully completed; other parts are attempted and their intents addressed, but they are not successfully completed. _____ There is evidence that the student has partial understanding of key concepts and procedures. _____ Some student work and explanations are clear, but it is necessary to make inferences to understand the response. _____ Additional illustrations or information, if present, may not enhance communication significantly. _____ Answers for some parts are correct, but partially correct or incorrect for others.
1	**Minimally accurate, complete, and clear** _____ A part (or parts) of the task is (are) addressed with minimal success, while other parts are omitted or incorrect. _____ There is minimal or limited evidence that the student understands concepts and procedures. _____ Student work and explanations may be difficult to follow, and it is necessary to fill in the gaps to understand the response. _____ Additional illustrations or information, if present, do not enhance communication and may be irrelevant. _____ Answers to most parts are incorrect.
0	**Not accurate, complete, and clear** _____ No part of the task is completed with any success. _____ There is little, if any, evidence that the student understands key concepts and procedures. _____ Student work and explanations are very difficult to follow and may be incomprehensible. _____ Any additional illustrations, if present, do not enhance communication and are irrelevant. _____ Answers to all parts are incorrect.

Performance Assessment

HARCOURT MATH Grade 2

Class Record Form

School	Unit 1		Unit 2		Unit 3		Unit 4	
Teacher	Task A	Task B	Task A	Task B	Task A	Task B	Task A	Task B
NAMES Date								

Performance Assessment

Class Record Form

School	Unit 5		Unit 6	
Teacher	Task A	Task B	Task A	Task B
NAMES Date				

TASK A

Planting Flowers

Purpose
To assess ability to write addends that show a combination equaling 12

Time
10–15 minutes

Grouping
Individuals or partners

Introduce the Task
Children are asked to chose flowers from the picture and write the addends that show a combination equaling 12. Tell them they may choose any combination of flowers as long as they spend $12.00.

TASK B

Veggie Soup

Purpose
To assess ability to choose a number to use in a group based on 2 parameters

Time
10–15 minutes

Grouping
Individuals or partners

Materials (optional)
Counters

Preparation Hints
You may wish to have children use manipulatives, such as different types of counters to represent the different vegetables.

Introduce the Task
Read through the task with the children. Tell them they must use a total of 15 vegetables, but they may use any combination of the vegetables shown in the picture to arrive at that number. They may only use the numbers of vegetables shown. For example, they may not use more than 1 onion.

Name ______________________ Date ______________

TASK A

Planting Flowers

Performance Indicators	Observations and Rubric Score
_______ Finds the sum of 3 or more addends. _______ Picks a combination that adds to $12. _______ Chooses addends that are $2, $3, or $4.	3 2 1 0

TASK B

Veggie Soup

Performance Indicators	Observations and Rubric Score
_______ Uses a total of 15 vegetables. _______ Uses at least one of each type of vegetable. _______ Shows 7 vegetables left over.	3 2 1 0

Total Score _______ /6

Name ____________________________

Planting Flowers

Juan has $12 to spend on flowers.

Show 2 ways Juan could spend $12 on flowers.

Show your work.

How many?

____

____

____

How many?

____

____

____

Name ______________________________

Veggie Soup

Paul wants to use at least one of each type of vegetable to make his soup. Write how many of each vegetable Paul will use if he chooses a total of 15 vegetables.

______ ______ ______ ______

Show your work.

Name ____________

UNIT 1 • TASK A

Planting Flowers

Juan has $12 to spend on flowers.

Show 2 ways Juan could spend $12 on flowers.

Show your work.

Possible combinations:

$4 + $4 + $4

$4 + $4 + $2 + $2

$4 + $3 + $3 + $2

$3 + $3 + $3 + $3

$3 + $3 + $2 + $2 + $2

$2 + $2 + $2 + $2 + $2 + $2

How many?

How many?

Performance Assessment PA3

Name ____________

UNIT 1 • TASK B

Veggie Soup

Paul wants to use at least one of each type of vegetable to make his soup. Write how many of each vegetable Paul will use if he chooses a total of 15 vegetables.

Show your work.

Possible answer:

Paul used 1 onion, 4 tomatoes, 6 ears of corn, and 4 potatoes.

He had no onions, no tomatoes, 3 ears of corn, and 4 potatoes left over.

(Any combination that adds to 15, uses all four types of vegetables and leaves 7)

PA4 Performance Assessment

Model Student Papers for

Planting Flowers

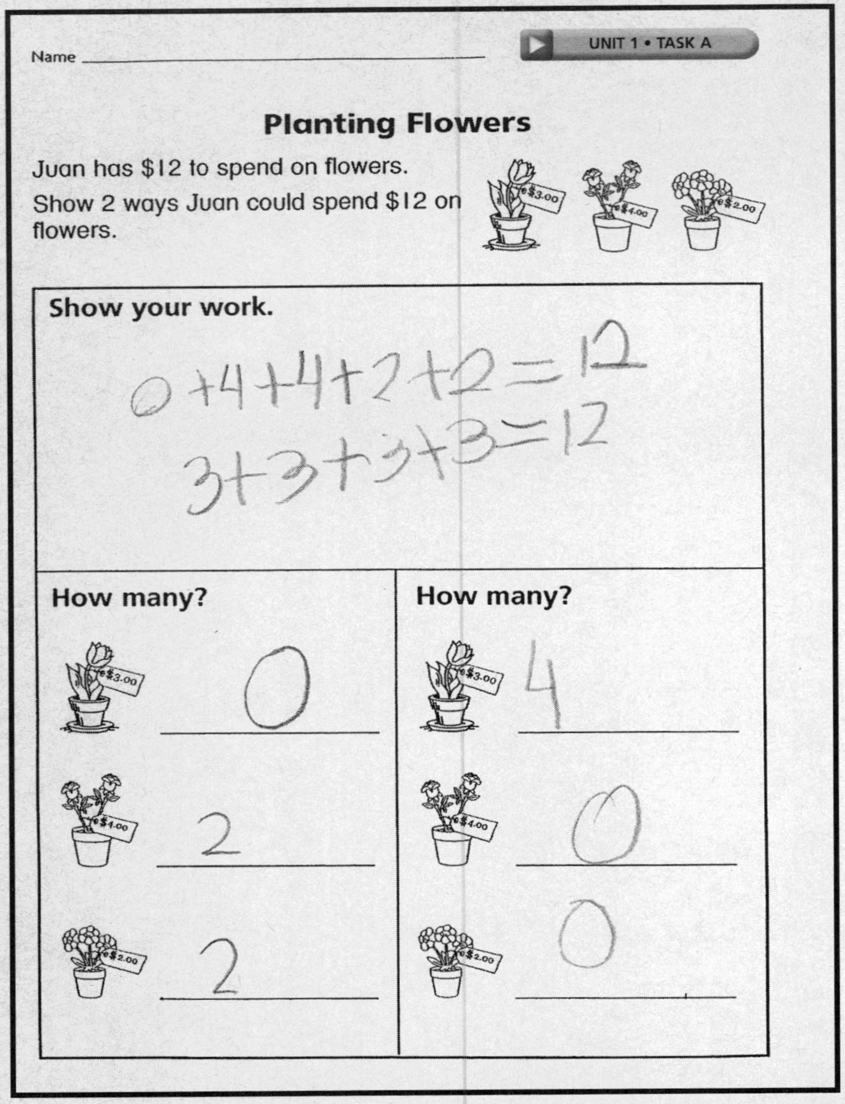
Name ______ UNIT 1 • TASK A

Planting Flowers

Juan has $12 to spend on flowers.
Show 2 ways Juan could spend $12 on flowers.

Show your work.

0+4+4+2+2=12
3+3+3+3=12

How many?		How many?	
$3.00	0	$3.00	4
$4.00	2	$4.00	0
$2.00	2	$2.00	0

Level 3 Child shows good understanding of the task. Child picks combinations that add up to $12.00, uses addends of $2, $3, or $4 and shows work clearly.

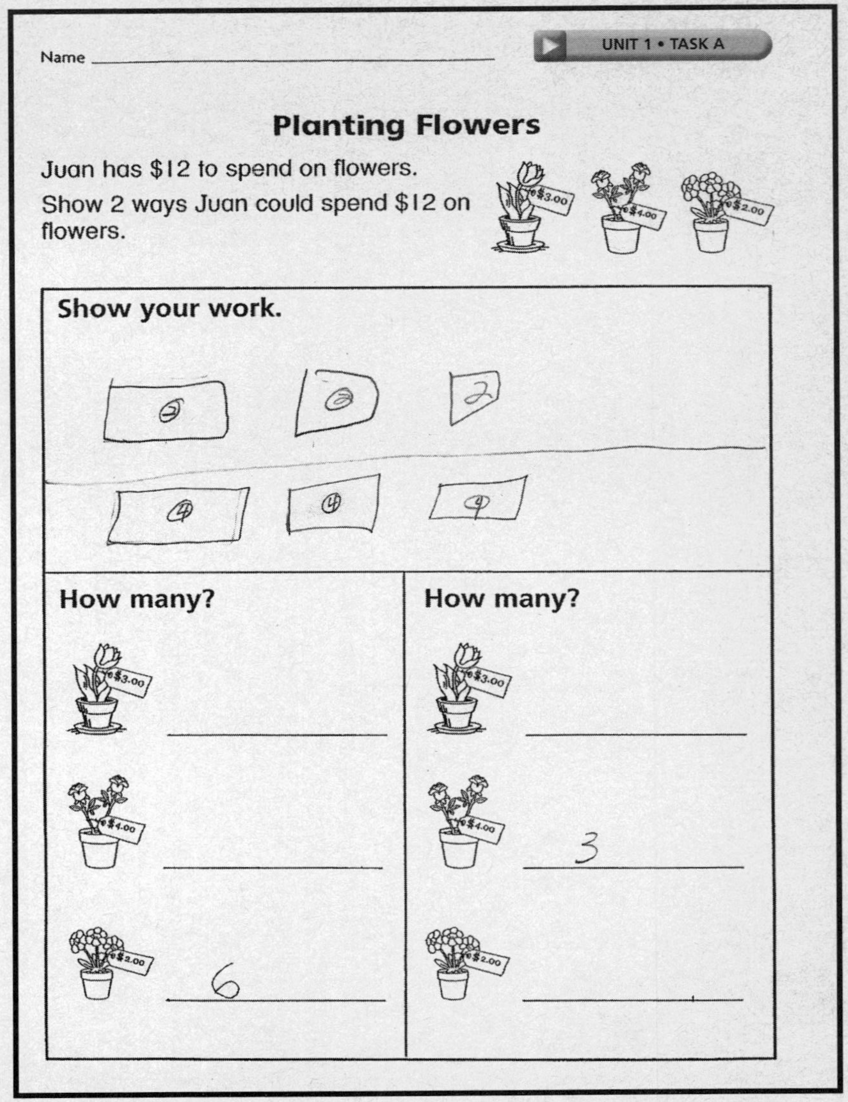
Name ______ UNIT 1 • TASK A

Planting Flowers

Juan has $12 to spend on flowers.
Show 2 ways Juan could spend $12 on flowers.

Show your work.

How many?		How many?	
$3.00		$3.00	
$4.00		$4.00	3
$2.00	6	$2.00	

Level 2 Child correctly completed the chart showing 2 ways to spend $12 but there is an error in the visual representation.

Model Student Papers for

Planting Flowers

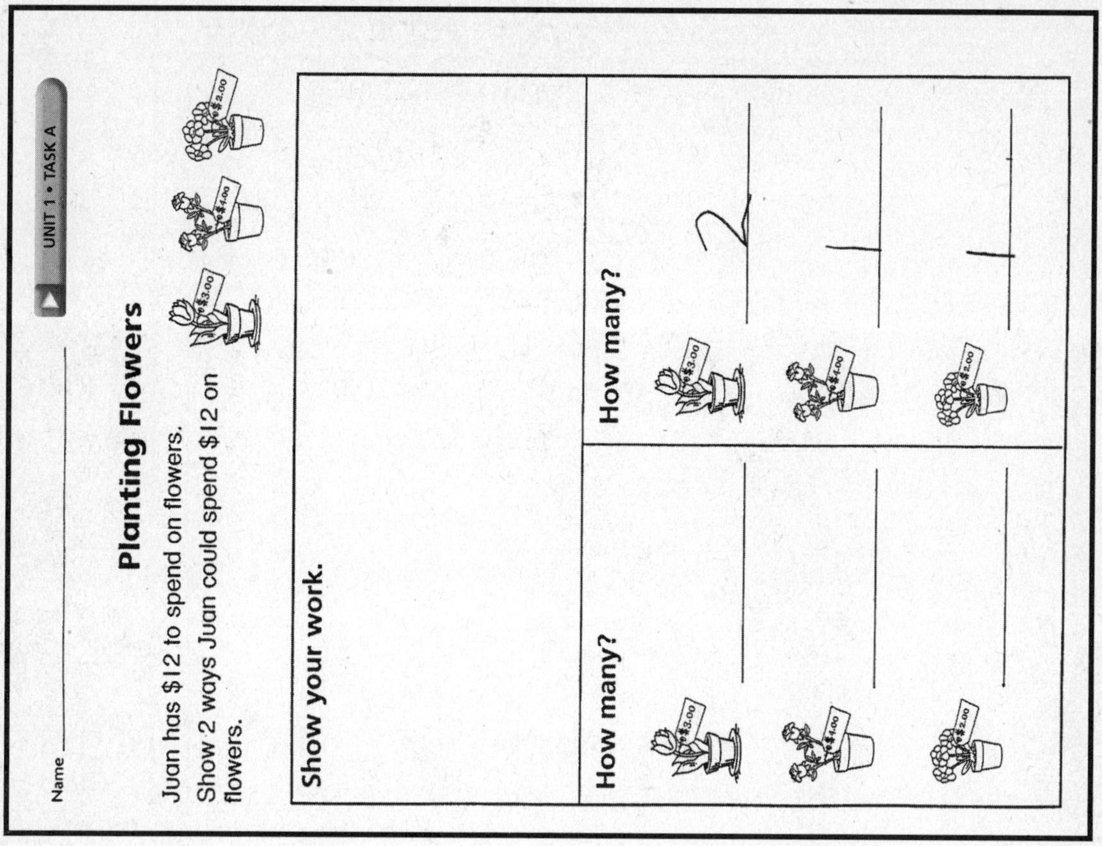

Name ____________

UNIT 1 • TASK A

Planting Flowers

Juan has $12 to spend on flowers.
Show 2 ways Juan could spend $12 on flowers.

$3.00 $4.00 $2.00

Show your work.

	How many?	How many?
$3.00		2
$4.00		1
$2.00		1

Level 1 The work as shown is correct but incomplete.

Veggie Soup

Name ______________________ UNIT 1 • TASK B

Veggie Soup

Paul wants to use at least one of each type of vegetable to make his soup. Write how many of each vegetable Paul can use if he chooses 15 vegetables.

8 1 5 1

Show your work.

8 + 1 + 5 + 1 = 15

Level 3 The child shows understanding of the task. Child writes a correct addition sentence to prove selection of the number of veggies to be correct.

Name ______________________ UNIT 1 • TASK B

Veggie Soup

Paul wants to use at least one of each type of vegetable to make his soup. Write how many of each vegetable Paul can use if he chooses 15 vegetables.

8 1 2 4

Show your work.

18 + 1 + 11 + 4 = 24

Level 2 The child is able to correctly choose the number of veggies to equal 15. There is some confusion in the attempt to show the addition sentence.

Model Student Papers for

Veggie Soup

Name ____________________

UNIT 1 • TASK B

Veggie Soup

Paul wants to use at least one of each type of vegetable to make his soup. Write how many of each vegetable Paul can use if he chooses 15 vegetables.

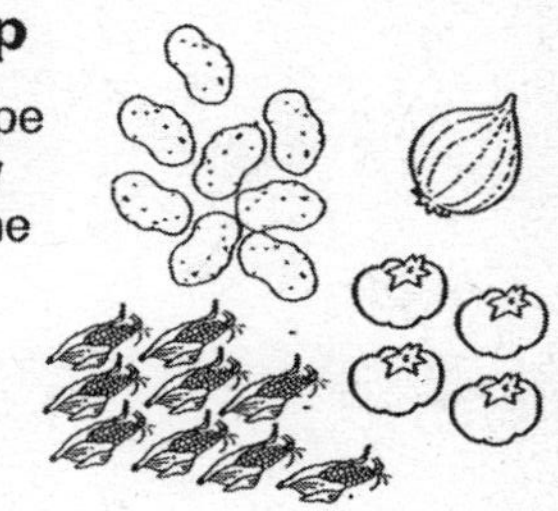

5   5   2   3

Show your work.

15
− 4

Level 1 The response shows confusion in the number of veggies that can be used. There is only 1 onion available but the child selected 5. The numbers do add up to 15.

TASK A

Shop for Toys

Purpose
To assess ability to use different methods to solve problems
To assess ability to add three numbers and estimate sums

Time
10–15 minutes

Grouping
Individuals or partners

Materials (optional)
Coins

Introduce the Task
Children are asked to solve the problem using estimation, mental math, and paper and pencil. They are asked to determine if they have enough money to buy all three toys.

TASK B

Box-Top Books

Purpose
To assess the ability to use add three 2-digit numbers with regrouping
To assess ability to subtract 2-digit numbers

Time
10–15 minutes

Grouping
Individuals or partners

Introduce the Task
Children are asked to solve the problem by first using addition, then subtraction. The children need to determine how many books a class has earned based on the number of box-tops collected. Then they need to determine how many more box-tops are needed to earn another book.

Name ______________________ Date ______________

TASK A

Shop for Toys

Performance Indicators	Observations and Rubric Score
_______ Estimates the sum of three toys to tell if they have enough money to buy them. _______ Explains how to use mental math to find the sum of three toys. _______ Use paper and pencil to find the sum of three toys.	3 2 1 0

TASK B

Box-Top Books

Performance Indicators	Observations and Rubric Score
_______ Adds three 2-digit numbers to find the total box tops collected. _______ Identifies how many books earned. _______ Subtracts to find how many more box tops are needed to earn another book.	3 2 1 0

Total Score _______ /6

Shop for Toys

Maggie has 87¢.

Explain how she can estimate to find out if she can buy all three items.

Explain how she can use mental math to find out if she can buy all three items.

Show how she can use paper and pencil to find out if she can buy all three items.

Box-Top Books

Second graders at Sussex School are collecting box tops. The school gets a new library book for every 50 box tops.

- Class 2-1 collected 44 box tops.
- Class 2-2 collected 15 box tops.
- Class 2-3 collected 39 box tops.

How many box tops did they collect altogether?

____

How many books did they earn so far? Explain your answer.

____

How many more box tops do they need for the next book? Explain your answer.

____

Show your work.

Name ______________________

UNIT 2 • TASK A

Shop for Toys

Maggie has 87¢.

Explain how she can estimate to find out if she can buy all three items.

Possible answer: She can round 31¢ to 30¢.
She can round 42¢ to 40¢.
She can add 30¢ + 20¢ + 40¢ = 90¢.
That's not enough money to buy all three toys.

Explain how she can use mental math to find out if she can buy all three items.

Check children's explanations.

Show how she can use paper and pencil to find out if she can buy all three items.

31¢
20¢
42¢
———
93¢

She does not have enough money.

PA12 Performance Assessment

Name ______________________

UNIT 2 • TASK B

Box-Top Books

Second graders at Sussex School are collecting box tops. The school gets a new library book for every 50 box tops.

- Class 2-1 collected 44 box tops.
- Class 2-2 collected 15 box tops.
- Class 2-3 collected 39 box tops.

How many box tops did they collect altogether?

98 box tops

How many books did they earn so far? Explain your answer.

1 book

How many more box tops do they need for the next book? Explain your answer.

2 box tops

Show your work.

$$\begin{array}{r}44\\15\\+\,39\\\hline 98\end{array}\qquad\begin{array}{r}98\\-\,50\\\hline 48\end{array}\qquad\begin{array}{r}50\\-\,48\\\hline 2\end{array}$$

Possible explanation:
They had 48 box tops left over after they got 1 book. Since they need 50 box tops for another book, I subtracted, 50 − 48 = 2.

Performance Assessment PA13

Model Student Papers for
Shop for Toys

Name ____________________ UNIT 2 • TASK A

Shop for Toys

Maggie has 87¢.

Explain how she can estimate to find out if she can buy all three items.

30
20
+40
90

She does not have enough to buy them all.

Explain how she can use mental math to find out if she can buy all three items.

31 30
20 30
42 +30
90

Show how she can use paper and pencil to find out if she can buy all three items.

42
31
20
93

Level 3 The child shows understanding of the task. The child estimates, uses mental math, and uses paper and pencil to find the sum of three 2-digit numbers.

Name ____________________ UNIT 2 • TASK A

Shop for Toys

Maggie has 87¢.

Explain how she can estimate to find out if she can buy all three items.

she cuold add in her head.

Explain how she can use mental math to find out if she can buy all three items.

Show how she can use paper and pencil to find out if she can buy all three items.

1 +0
31
+20
=42
93

Level 2 The child shows some understanding of the task, but did not complete all parts of the task.

Model Student Papers for
Shop for Toys

Name ____________________ UNIT 2 • TASK A

Shop for Toys

Maggie has 87¢.

Explain how she can estimate to find out if she can buy all three items.

Adding It. to see how much is It.

Explain how she can use mental math to find out if she can buy all three items.

to Add It up put the three unDer eke uthere

Show how she can use paper and pencil to find out if she can buy all three items.

If She had paper and a pencil Then She could Add It up.

Level 1 The child attempted all parts of the task, but did not do any of the computation or estimation.

Box-Top Books

Name ____________________

UNIT 2 • TASK B

Box-Top Books

Second graders at Sussex School are collecting box tops. The school gets a new library book for every 50 box tops.

- Class 2-1 collected 44 box tops.
- Class 2-2 collected 15 box tops.
- Class 2-3 collected 39 box tops.

How many box tops did they collect altogether?

98

How many books did they earn so far? Explain your answer.

1 Because they need 100 too get two.

How many more box tops do they need for the next book? Explain your answer.

2 Because 98+2 is 100.

Show your work.

10 + 44 + 15 + 39 = 98 ×8 40 +98 +2 = 100 they need 2

Level 3 The child shows understanding of the task. The child correctly finds the sum. The child correctly identifies how many books the children earned and how many more box tops are needed to earn another book.

Name ____________________

UNIT 2 • TASK B

Box-Top Books

Second graders at Sussex School are collecting box tops. The school gets a new library book for every 50 box tops.

- Class 2-1 collected 44 box tops.
- Class 2-2 collected 15 box tops.
- Class 2-3 collected 39 box tops.

How many box tops did they collect altogether?

98

How many books did they earn so far? Explain your answer.

1

How many more box tops do they need for the next book? Explain your answer.

Show your work.

44 +15 +39 = 98 ×8 48

Level 2 The child shows understanding of the task, however the child did not complete the task.

Model Student Papers for
Box-Top Books

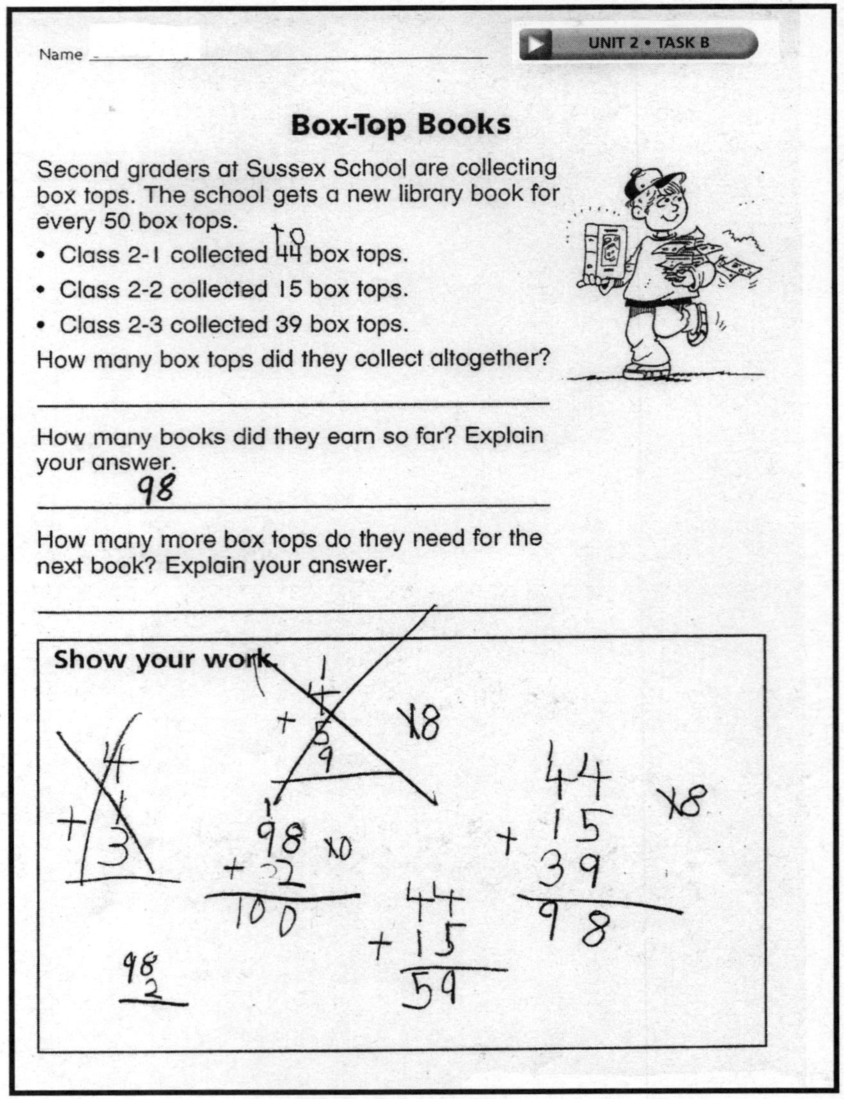

UNIT 2 • TASK B

Name

Box-Top Books

Second graders at Sussex School are collecting box tops. The school gets a new library book for every 50 box tops.

- Class 2-1 collected 44 box tops.
- Class 2-2 collected 15 box tops.
- Class 2-3 collected 39 box tops.

How many box tops did they collect altogether?

How many books did they earn so far? Explain your answer.

How many more box tops do they need for the next book? Explain your answer.

Show your work.

Level 1 The child correctly added the three numbers, but doesn't show understanding of the task.

TASK A

Spin a Color

Purpose
To assess ability to make a prediction based on a spinner
To assess ability to record data in a tally table

Time
10–15 minutes

Grouping
Individuals or partners

Materials
Paper clips, crayons, pencils

Introduce the Task
Children are asked to color a spinner and then make a prediction about the color on which the spinner will most likely land.

TASK B

Stuck on You

Purpose
To assess the ability to combine coins to total a given amount

Time
10–15 minutes

Grouping
Individuals or partners

Materials (optional)
Coins

Preparation Hints
Read through the task with the children. Review the value of coins.

Introduce the Task
Children are asked to choose the number of coins based on clues given. They are asked to find the change from 55¢ after the purchase of 2 stickers each costing 25¢. They make a model of various coin combinations.

Name ______________________ Date ______________

TASK A

Spin a Color

Performance Indicators	Observations and Rubric Score
______ Colors the spinner and predicts a color on which the pointer will most likely land. ______ Records data in the tally table for 20 spins. ______ Draws a spinner that will most likely land on green.	3 2 1 0

TASK B

Stuck on You

Performance Indicators	Observations and Rubric Score
______ Shows two combinations between 3 and 10 coins. ______ Coins total 55¢. ______ Shows correct change of 5¢.	3 2 1 0

Total Score ______ /6

Name ______________________________

Spin a Color

Use a pencil and paper clip to make a spinner.

Color the spinner. Color one part red, one part blue, and one part yellow.

Predict. On which color will the pointer most often stop?

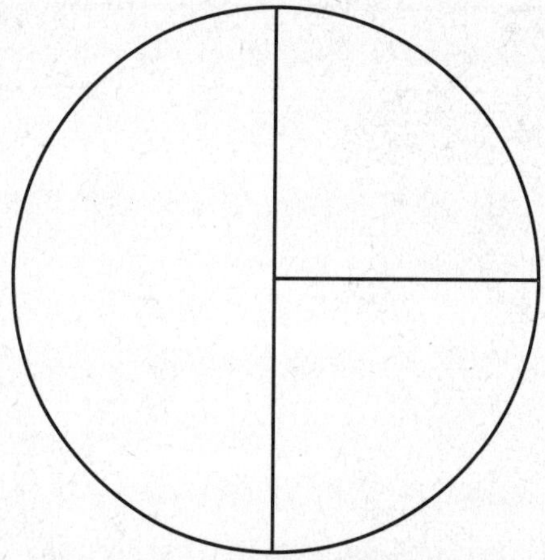

____

Try it. Spin the paper clip.

Spin 20 times.

Make a tally mark after each spin.

Color	Tally Marks	Total
red		
blue		
yellow		

On which color did the pointer most often stop? Explain why.

Draw a spinner that has three colors and that you think will most often stop on green.

Name ____________________

Stuck on You

Jane wants to buy these 2 stickers.

- She has 55¢ in her pocket.
- She has more than 3 coins, but less than 10 coins.

Show 2 ways Jane can have the coins in her pocket.

How much money will Jane have left after she buys the stickers?

Show your work.

Name ____________________

UNIT 3 • TASK A

Spin a Color

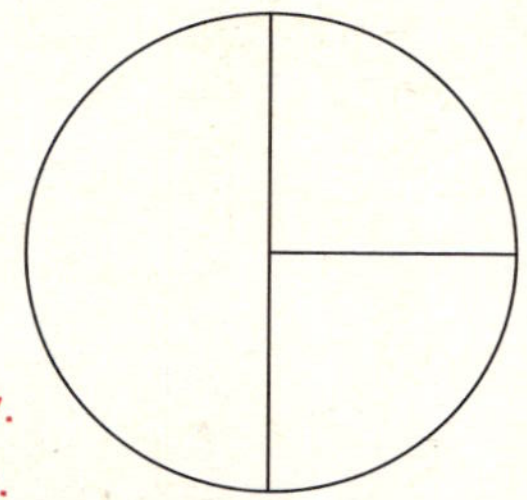

Use a pencil and paper clip to make a spinner.

Color the spinner. Color one part red, one part blue, and one part yellow.

Predict. On which color will the pointer most often stop? Answers will vary.

Check children's spinners and predictions.

Try it. Spin the paper clip.

Spin 20 times.

Make a tally mark after each spin.

On which color did the pointer most often stop? Explain why.

Color	Tally Marks	Total
red		
blue		
yellow		

Check children's tally tables for 20 spins. Answers will vary depending on data.

Draw a spinner that has three colors and that you think will most often stop on green.

Check children's spinners.

Performance Assessment PA21

Name ____________________

UNIT 3 • TASK B

Stuck on You

Jane wants to buy these 2 stickers.

- She has 55¢ in her pocket.
- She has more than 3 coins, but less than 10 coins.

Show 2 ways Jane can have the coins in her pocket.

How much money will Jane have left after she buys the stickers?

Show your work.

Possible answers:

Jane had 5 coins: 1 quarter, 2 dimes, and 2 nickels totaling 55¢. If she bought the stickers she would spend 50¢, with 5¢ left over.

Other combinations include 5 dimes and 1 nickel, 4 dimes and 3 nickels; 3 dimes and 5 nickels; 2 dimes, 7 nickels; 1 quarter, 3 dimes; 1 quarter, 1 dime, 4 nickels.

PA22 Performance Assessment

Spin a Color

Name ______________________

UNIT 2 • TASK A

Spin a Color

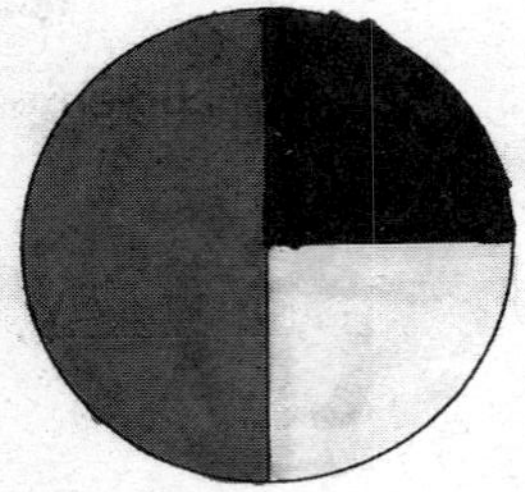

Use a pencil and paper clip to make a spinner.

Color the spinner. Color one part red, one part blue, and one part yellow.

Predict. On which color will the pointer most often stop?

red

Try it. Spin the paper clip.

Spin 20 times.

Make a tally mark after each spin.

Color	Tally Marks	Total
red		13
blue		3
yellow		4

On which color did the pointer most often stop? Explain why.

red couse it was the biggest space.

Draw a spinner that has three colors and that you think will most often stop on green.

Level 3 The child shows understanding of the task. The child completes the tally table, gives an explanation of the outcomes and correctly draws another spinner that will most likely land on green.

Name ______________________

UNIT 3 • TASK A

Spin a Color

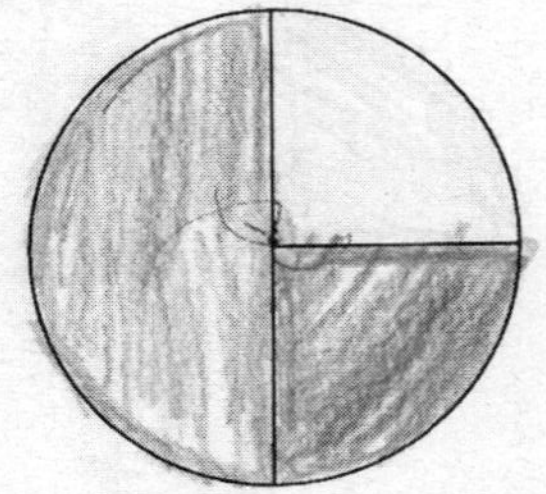

Use a pencil and paper clip to make a spinner.

Color the spinner. Color one part red, one part blue, and one part yellow.

Predict. On which color will the pointer most often stop?

Blue

Try it. Spin the paper clip.

Spin 20 times.

Make a tally mark after each spin.

Color	Tally Marks	Total
red		3
blue		16
yellow		5

On which color did the pointer most often stop? Explain why.

Blue

Because it was in the Big spas.

Draw a spinner that has three colors and that you think will most often stop on green.

Level 2 The child shows understanding of the task, but has some errors in the tally table. Other parts of the task are correct.

Spin a Color

Name ______________________

UNIT 3 • TASK A

Spin a Color

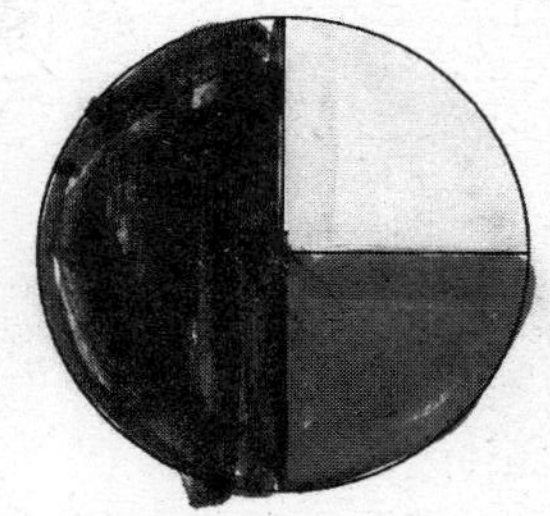

Use a pencil and paper clip to make a spinner.

Color the spinner. Color one part red, one part blue, and one part yellow.

Predict. On which color will the pointer most often stop?

Blue

Try it. Spin the paper clip.

Spin 20 times.

Make a tally mark after each spin.

Color	Tally Marks	Total
red	||	2
blue	+||||||||||||	13
yellow	|||||||	7

On which color did the pointer most often stop? Explain why.

Blue was the most because I Put blue in the big spase.

Draw a spinner that has three colors and that you think will most often stop on green.

Level 1 The child shows some understanding, but there are errors in the tally table and the new spinner was drawn incorrectly.

Model Student Papers for

Stuck on You

Name ____________________

UNIT 3 • TASK B

Stuck on You

Jane wants to buy these 2 stickers.

- She has 55¢ in her pocket.
- She has more than 3 coins, but less than 10 coins.

Show 2 ways Jane can have the coins in her pocket.

How much money will Jane have left after she buys the stickers?

Show your work.

25 25 1 1 1 1

1 = 55¢

25 10 10 10 = 55¢

55¢
- 50
05   05¢

Level 3 Response shows understanding of problem solving and ability to perform multi-step functions. The child shows 2 ways to make 55¢, meeting the task requirements. The child correctly uses subtraction.

Name ____________________

UNIT 3 • TASK B

Stuck on You

Jane wants to buy these 2 stickers.

- She has 55¢ in her pocket.
- She has more than 3 coins, but less than 10 coins.

Show 2 ways Jane can have the coins in her pocket.

How much money will Jane have left after she buys the stickers?

Show your work.

10 10 10 10 10 5 = 55¢

5 5 10 10 10 10 5 = 55¢

Level 2 The child correctly shows 2 ways to make 55¢, but fails to show how much money is left after buying the stickers.

Name ______________________ UNIT 3 • TASK B

Stuck on You

Jane wants to buy these 2 stickers.

- She has 55¢ in her pocket.
- She has more than 3 coins, but less than 10 coins.

Show 2 ways Jane can have the coins in her pocket.

How much money will Jane have left after she buys the stickers?

Show your work.

Level 1 The child demonstrated one way to make 55¢ but did not show ability to understand multi-step process.

TASK A

Find Shapes

Purpose
To assess ability to identify solid figures
To assess ability to find objects that have the same shape as solid figures provided

Time
10–15 minutes

Grouping
Individuals or partners

Materials
solid figures: rectangular prism, sphere, square pyramid, cylinder; several objects of each shape; crayons and pencils

Introduce the Task
Children are asked to find objects in the classroom that have the same shape as the blocks on the page. Review the names of the shapes and their faces with the children.

TASK B

Door Pattern

Purpose
To assess ability to create a pattern
To assess ability to identify congruent figures
To assess ability to identify flips and turns

Time
10–15 minutes

Grouping
Individuals or partners

Materials
Pattern blocks or construction paper, crayons, pencils

Introduce the Task
Children are asked to create a colorful pattern to decorate a door. The pattern must use congruent shapes and show flips and turns of the shapes.

Name ____________________ Date ____________

TASK A

Find Shapes

Performance Indicators	Observations and Rubric Score
_______ Identifies objects in the classroom that are the same shape as the solid figures. _______ Draws pictures of the classroom objects in the box with the matching solid figure. _______ Explains how they know the objects are the same shape.	3 2 1 0

TASK B

Door Pattern

Performance Indicators	Observations and Rubric Score
_______ Makes a pattern. _______ Uses flips and turns. _______ Uses congruent shapes.	3 2 1 0

Total Score _______ /6

Name ______________________

Find Shapes

Look at the solid figures.

Draw pictures of objects in the classroom that are the same shape.

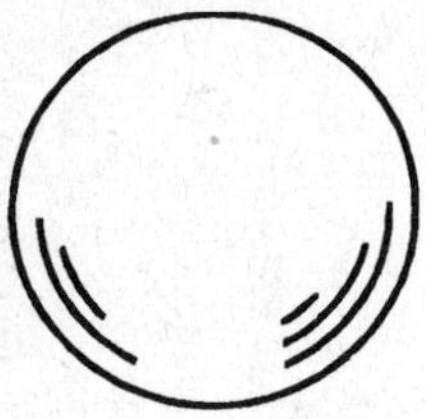

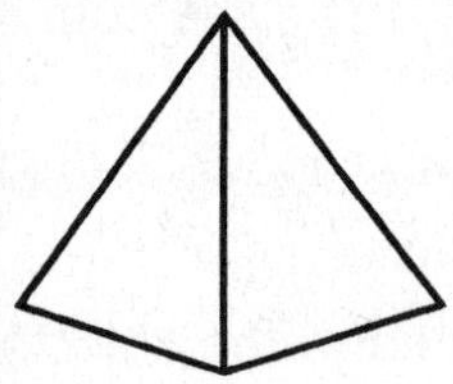

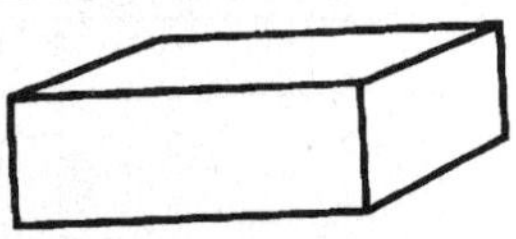

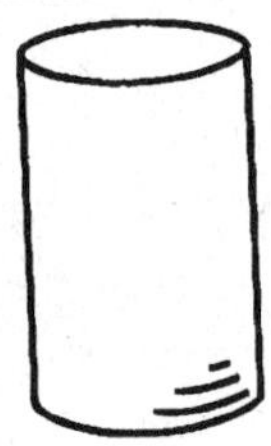

Explain how you know when objects are the same shape.

Name ______________________________

Door Pattern

Amy wants to decorate her bedroom door with shape stickers. First she plans the pattern on a sheet of paper.

- The pattern uses fun colors.
- The pattern uses shapes with flips and turns.
- The pattern uses congruent shapes.

Draw one way the pattern could look.

Show your work.

Name ______________________

UNIT 4 • TASK A

Find Shapes

Look at the solid figures.

Draw pictures of objects in the classroom that are the same shape.

Check children's drawings.

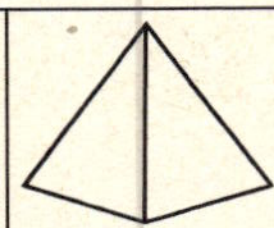

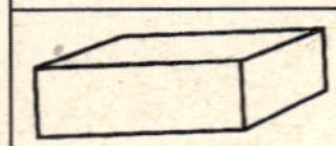

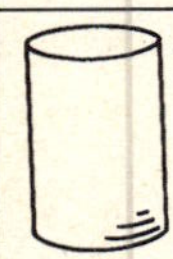

Explain how you know when objects are the same shape.

Check children's explanations.

Name ______________________

UNIT 4 • TASK B

Door Pattern

Amy wants to decorate her bedroom door with shape stickers. First she plans the pattern on a sheet of paper.

- The pattern uses fun colors.
- The pattern uses shapes with flips and turns.
- The pattern uses congruent shapes.

Draw one way the pattern could look.

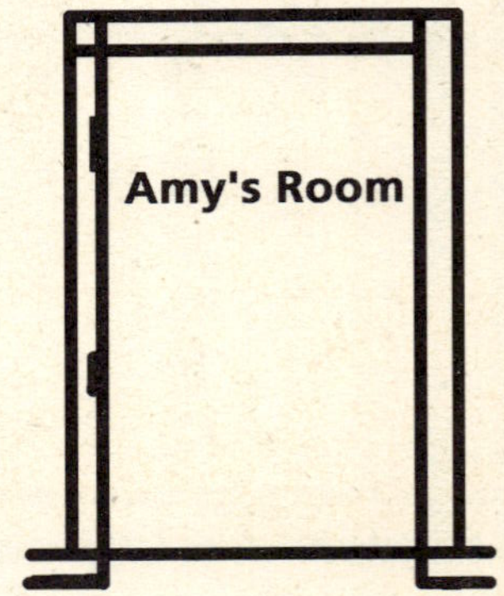

Show your work.

Check children's patterns.

Find Shapes

Name ______________________

UNIT 4 • TASK A

Find Shapes

Look at the solid figures.

Draw pictures of objects in the classroom that are the same shape.

Explain how you know when objects are the same shape.

you look a the sides.

Level 3 The child shows understanding of the task. The child identifies and draws objects in the classroom that are the same as the solid figures. An explanation is given.

Name ______________________

UNIT 4 • TASK A

Find Shapes

Look at the solid figures.

Draw pictures of objects in the classroom that are the same shape.

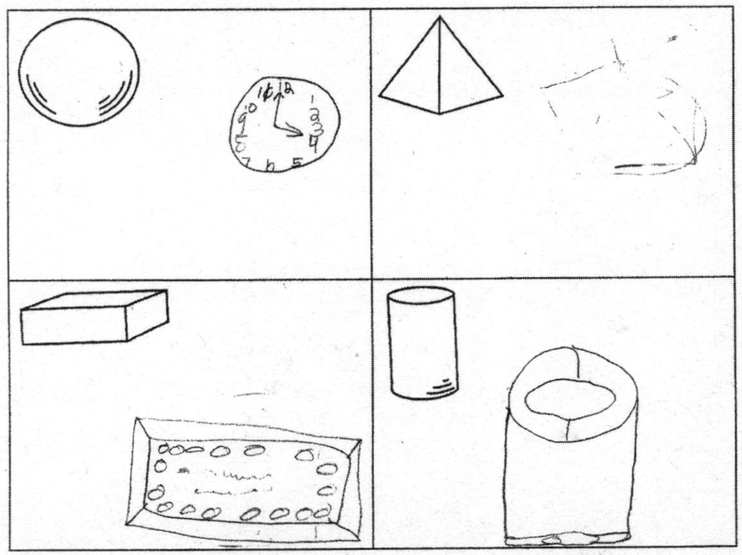

Explain how you know when objects are the same shape.

Level 2 The child shows some understanding of the task. The child correctly identified classroom objects for two of the four solid figures.

Model Student Papers for
Find Shapes

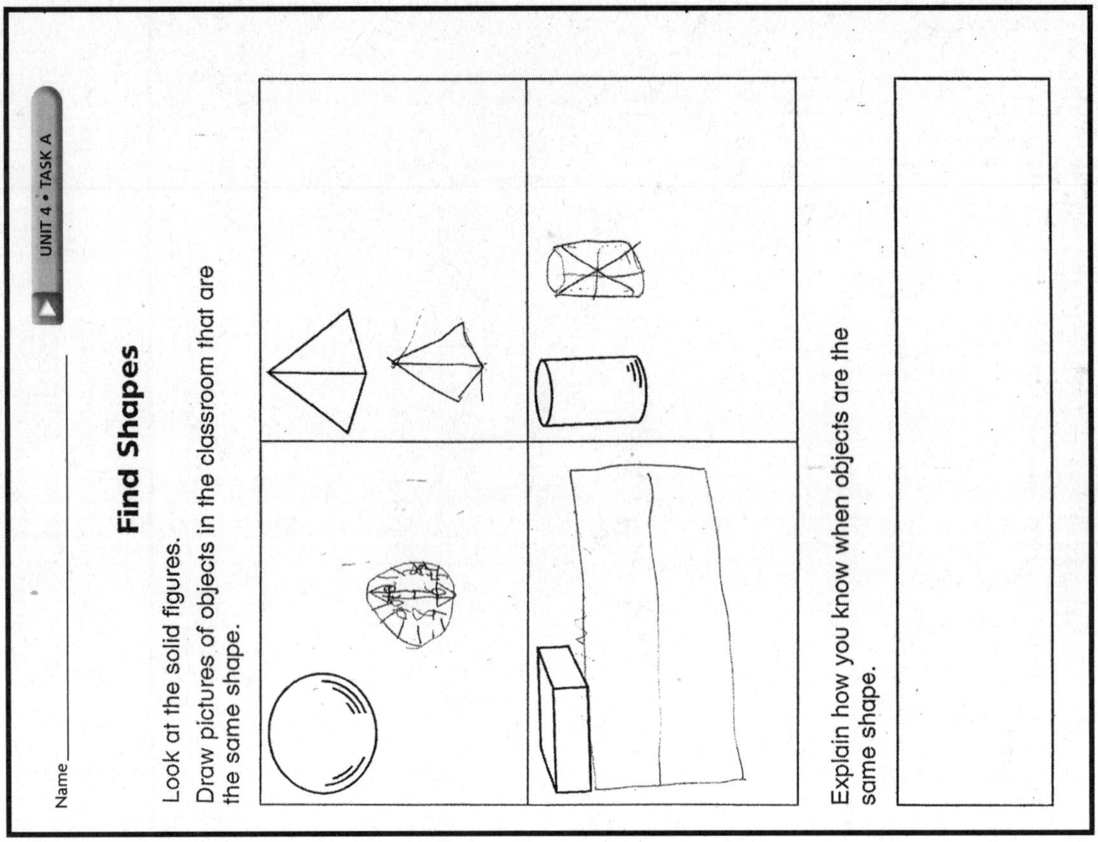

Name ________

UNIT 4 • TASK A

Find Shapes

Look at the solid figures.

Draw pictures of objects in the classroom that are the same shape.

Explain how you know when objects are the same shape.

Level 1 The child shows limited understanding of the task. A cylinder was drawn, but crossed out. The clock was incorrectly identified as a sphere.

Door Pattern

Name ____________________ UNIT 4 • TASK B

Door Pattern

Amy wants to decorate her bedroom door with shape stickers. First she plans the pattern on a sheet of paper.

- The pattern uses fun colors.
- The pattern uses shapes with flips and turns.
- The pattern uses congruent shapes.

Draw one way the pattern could look.

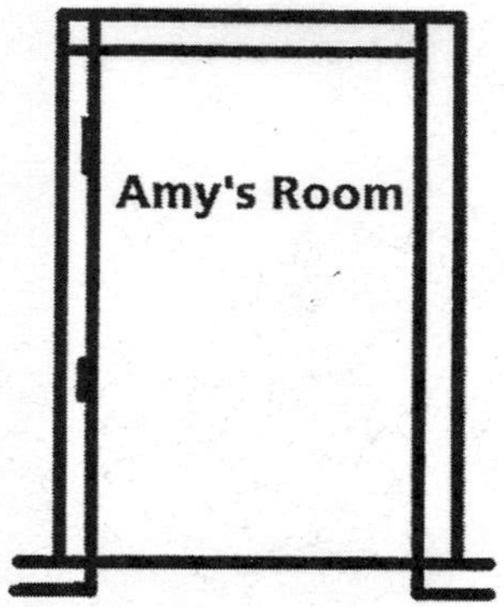

Show your work.

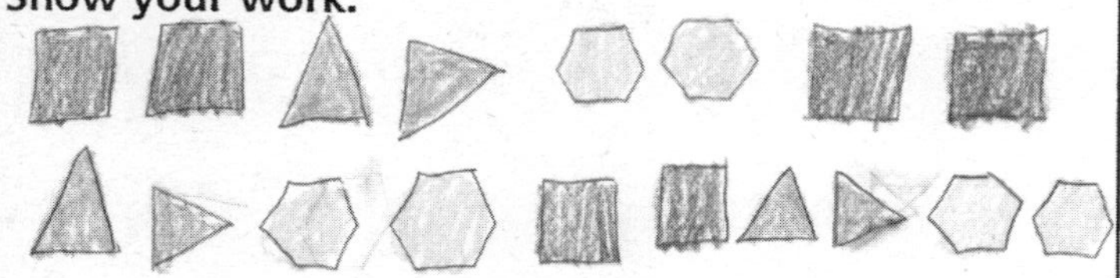

Level 3 The child shows understanding of the task. The child makes a pattern, shows flips and turns, and uses congruent shapes.

Name ____________________ UNIT 4 • TASK B

Door Pattern

Amy wants to decorate her bedroom door with shape stickers. First she plans the pattern on a sheet of paper.

- The pattern uses fun colors.
- The pattern uses shapes with flips and turns.
- The pattern uses congruent shapes.

Draw one way the pattern could look.

Show your work.

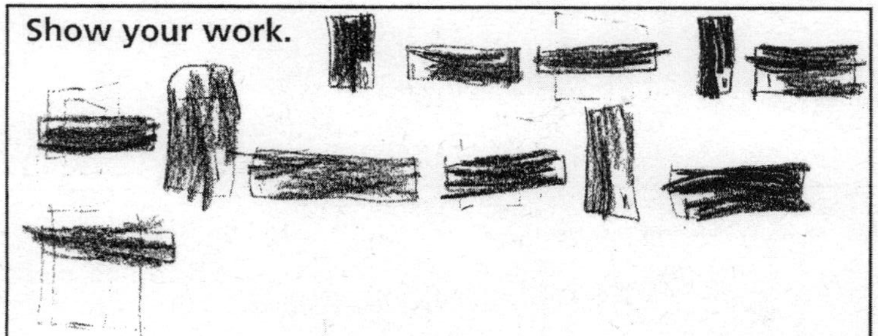

Level 2 The child makes a pattern using congruent shapes. The pattern is correct with one shape standing, 2 shapes turned on the side. The color pattern is inconsistent.

Door Pattern

Name ______________________

UNIT 4 • TASK B

Door Pattern

Amy wants to decorate her bedroom door with shape stickers. First she plans the pattern on a sheet of paper.

- The pattern uses fun colors.
- The pattern uses shapes with flips and turns.
- The pattern uses congruent shapes.

Draw one way the pattern could look.

Show your work.

Level 1 The child attempts to make a color pattern, but uses all the same shapes and indicates no flips and turns.

TASK A

Frame the Picture

Purpose
To assess ability to use an inch ruler to measure length
To assess ability to find perimeter

Time
10–15 minutes

Grouping
Individuals or partners

Materials
Inch ruler, yarn (optional)

Introduce the Task
Children are asked to measure the distance around a picture so they can determine the amount of yarn needed to frame it.

TASK B

Hats Off!

Purpose
To assess the ability to solve problems by making a model
To assess the ability to identify parts of a group by using fractions
To assess the ability to identify fractions as parts of a whole

Time
10–15 minutes

Grouping
Individuals or partners

Preparation Hints
Read through the task with the children. You may wish to have children use counters or other manipulatives to represent the players.

Introduce the Task
Children are asked to choose the number of players in a group who are wearing helmets.

Name ______________________________ Date ______________

TASK A

Frame the Picture

Performance Indicators	Observations and Rubric Score
_______ Measure the distance around the picture. _______ Explains how much yarn is needed.	3 2 1 0

TASK B

Hats Off!

Performance Indicators	Observations and Rubric Score
_______ Draws a group of more than 7 players. _______ Circles one-third of the group drawn. _______ Shows the fraction as part of the group.	3 2 1 0

Total Score _______ /6

Name ___________________________________

Frame the Picture

Renee is going to use yarn to make a frame for her picture. How much yarn does she need? Use an inch ruler to measure.

____

Explain how you decided how much yarn Renee needs.

Name ______________________________________

Hats Off!

Wendy saw a group of football players standing on the sideline.

- One-third of the players were wearing helmets.
- There were more than 7 players in the group.

Draw a picture to show the possible number of players in the group. Circle the number of players in the group who were wearing helmets.

Show your work.

Name ________________

UNIT 5 • TASK A

Frame the Picture

Renee is going to use yarn to make a frame for her picture. How much yarn does she need? Use an inch ruler to measure.

22 inches

Explain how you decided how much yarn Renee needs.

Possible answer:
I measured each side and added the numbers.
5 + 6 + 5 + 6 = 22 inches.

Performance Assessment PA39

Name ________________

UNIT 5 • TASK B

Hats Off!

Wendy saw a group of football players standing on the sideline.

- One-third of the players were wearing helmets.
- There were more than 7 players in the group.

Draw a picture to show the possible number of players in the group. Circle the number of players in the group who were wearing helmets.

Show your work.

Possible answer:

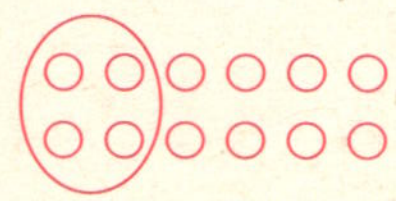

PA40 Performance Assessment

Frame the Picture

Name ____________

UNIT 5 • TASK A

Frame the Picture

Renee is going to use yarn to make a frame for her picture. How much yarn does she need? Use an inch ruler to measure.

____

Explain how you decided how much yarn Renee needs.

22 in. I measured 6 on the sids and 5 on the top and boutum.

Level 3 The child shows understanding of the task. The child correctly measures around the picture and identifies the distance. The child gives an explanation of the method.

Name ____________

UNIT 5 • TASK A

Frame the Picture

Renee is going to use yarn to make a frame for her picture. How much yarn does she need? Use an inch ruler to measure.

26

Explain how you decided how much yarn Renee needs.

7+7+6+6=26 you add 6+6+7+7 and count up.

Level 2 The child shows understanding of the task and provides an explanation of the method. However, the child measures incorrectly.

Frame the Picture

Name ______________________

UNIT 5 • TASK A

Frame the Picture

Renee is going to use yarn to make a frame for her picture. How much yarn does she need? Use an inch ruler to measure.

Explain how you decided how much yarn Renee needs.

13 16 13 16

Level 1 The child attempted the task and measured the length on each side. However, the measurements are incorrect and the total was not given.

Hats Off!

Name ____________________

UNIT 5 • TASK B

Hats Off!

Wendy saw a group of football players standing on the sideline.

- One-third of the players were wearing helmets.
- There were more than 7 players in the group.

Draw a picture to show the possible number of players in the group. Circle the number of players in the group who were wearing helmets.

Show your work.

Level 3 The child shows complete understanding of the task. Child draws a picture with a group greater than 7 and circles $\frac{1}{3}$ of the group shown wearing helmets. Uses a correct fraction.

Name ____________________

UNIT 5 • TASK B

Hats Off!

Wendy saw a group of football players standing on the sideline.

- One-third of the players were wearing helmets.
- There were more than 7 players in the group.

Draw a picture to show the possible number of players in the group. Circle the number of players in the group who were wearing helmets.

Show your work.

Level 2 The child solved the problem by making a model and used a fraction, but was not specific enough to show $\frac{1}{3}$ of the whole group.

Name ____________________

UNIT 5 • TASK B

Hats Off!

Wendy saw a group of football players standing on the sideline.

- One-third of the players were wearing helmets.
- There were more than 7 players in the group.

Draw a picture to show the possible number of players in the group. Circle the number of players in the group who were wearing helmets.

Show your work.

Level 1 The response shows partial understanding. The child shows $\frac{1}{3}$ of the group not wearing helmets, but does not show a group of more than 7.

TASK A

Dog Toys

Purpose
To assess ability to add and subtract money amounts
To assess ability to solve a multi-step problem

Time
10–15 minutes

Grouping
Individuals or partners

Materials (optional)
You may wish to provide dollar bills and coins for the children to use.

Introduce the Task
Children are asked to choose 2 toys from a list and compute their total cost. Then they find the amount of money Kyle had left after buying the toys.

TASK B

Here Kitty, Kitty

Purpose
To assess ability to multiply with 5
To assess ability to solve problems by choosing the operation

Time
10–15 minutes

Grouping
Individuals or partners

Preparation Hints
You may wish to have children use counters or other manipulatives to represent the kittens.

Introduce the Task
Children are asked to determine the number of kittens in a group, based on clues given. The total number of kittens must be greater than 18.

Name ______________________ Date ______________

TASK A Dog Toys

Performance Indicators	Observations and Rubric Score
_______ Picks two toys to buy and finds the correct sum for their cost. _______ Finds the correct change. _______ Has more than $1.00 left.	3 2 1 0

TASK B Here Kitty, Kitty

Performance Indicators	Observations and Rubric Score
_______ Picks a multiple of 5. _______ Picks a number greater than 18. _______ Shows correct multiplication and addition sentences.	3 2 1 0

Total Score _______ /6

Name ______________________________

Dog Toys

Kyle went to the pet store to pick out dog toys for his dog, Duke. Here are the toys he had to choose from:

Kyle had $5.75 to buy toys. He bought 2 toys.

Kyle had more than $1.00 left after he bought the toys.

Which 2 toys could Kyle have bought? How much money did he have left?

Show your work.

Name ____________________

Here Kitty, Kitty

Allan saw some kittens at the pet store. He decided to find out how many there were in all.

The kittens were in groups of 5. There were more than 18 kittens in the store.

Use both addition and multiplication to show how many kittens could have been in the store.

Show your work.

Name ____________________

UNIT 6 • TASK A

Dog Toys

Kyle went to the pet store to pick out dog toys for his dog, Duke. Here are the toys he had to choose from:

Kyle had \$5.75 to buy toys. He bought 2 toys.

Kyle had more than \$1.00 left after he bought the toys.

Which 2 toys could Kyle have bought? How much money did he have left?

Show your work.

Possible answer:

Kyle bought 1 little bone for \$1.33 and 1 ball for \$2.25, for a total of \$3.58; he had \$2.17 left. Other answers are possible.

Name ____________________

UNIT 6 • TASK B

Here Kitty, Kitty

Allan saw some kittens at the pet store. He decided to find out how many there were in all.

The kittens were in groups of 5. There were more than 18 kittens in the store.

Use both addition and multiplication to show how many kittens could have been in the store.

Show your work.

Possible answer:

25 kittens could have been in the store.

5 x 5 = 25, which is over 18.

5 + 5 + 5 + 5 + 5 = 25

Model Student Papers for
Dog Toys

Name ____________ UNIT 6 • TASK A

Dog Toys

Kyle went to the pet store to pick out dog toys for his dog, Duke. Here are the toys he had to choose from:

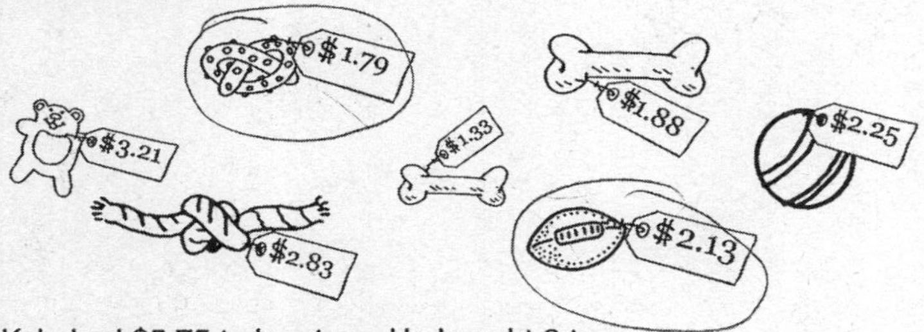

Kyle had $5.75 to buy toys. He bought 2 toys.

Kyle had more than $1.00 left after he bought the toys.

Which 2 toys could Kyle have bought? How much money did he have left?

Show your work.

$1.79 + 2.13 = $3.92

5.75 − 3.92 = $1.83

he had this buch left

Level 3 Response demonstrates ability to solve multi-step problems. The child shows understanding of adding and subtracting money. The child chooses two toys, finds the correct sum, and finds the amount left over.

Name ____________ UNIT 6 • TASK A

Dog Toys

Kyle went to the pet store to pick out dog toys for his dog, Duke. Here are the toys he had to choose from:

Kyle had $5.75 to buy toys. He bought 2 toys.

Kyle had more than $1.00 left after he bought the toys.

Which 2 toys could Kyle have bought? How much money did he have left? 1.58

ball + foot ball

Show your work.

2.25 + 2.13 = $4.38

Level 2 The child chooses two toys and performs the addition function correctly. The subtraction step is not shown and the answer is not correct.

Dog Toys

Name ______________________

UNIT 6 • TASK A

Dog Toys

Kyle went to the pet store to pick out dog toys for his dog, Duke. Here are the toys he had to choose from:

Kyle had $5.75 to buy toys. He bought 2 toys.

Kyle had more than $1.00 left after he bought the toys.

Which 2 toys could Kyle have bought? How much money did he have left?

Show your work.

He can buy the teddybear

He would have $2.54

$$
\begin{array}{r} \$5.75 \\ -\ 3.21 \\ \hline 2.54 \end{array}
$$

Level 1 Work shown is correct, but incomplete. The child did not choose two toys. The addition step is not shown, but the subtraction is correct.

Model Student Papers for
Here Kitty, Kitty

Name ______________________ UNIT 6 • TASK B

Here Kitty, Kitty

Allan saw some kittens at the pet store. He decided to find out how many there were in all.

The kittens were in groups of 5. There were more than 18 kittens in the store.

Use both addition and multiplication to show how many kittens could have been in the store.

Show your work.

4 × 5 = 20

5 + 5 + 5 + 5 = 20

Level 3 The child followed the parameters of the task. Child used multiples of five and showed both correct addition and multiplication sentences.

Name ______________________ UNIT 6 • TASK B

Here Kitty, Kitty

Allan saw some kittens at the pet store. He decided to find out how many there were in all.

The kittens were in groups of 5. There were more than 18 kittens in the store.

Use both addition and multiplication to show how many kittens could have been in the store.

Show your work.

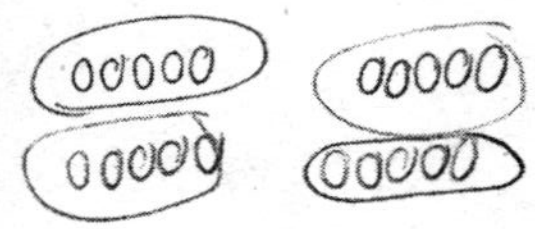

5 × 4 = 20

Level 2 The work shown is correct, but incomplete. The multiplication sentence is correct, but the addition sentence is not included.

Name ____________________ UNIT 6 • TASK B

Here Kitty, Kitty

Allan saw some kittens at the pet store. He decided to find out how many there were in all.

The kittens were in groups of 5. There were more than 18 kittens in the store.

Use both addition and multiplication to show how many kittens could have been in the store.

Show your work.

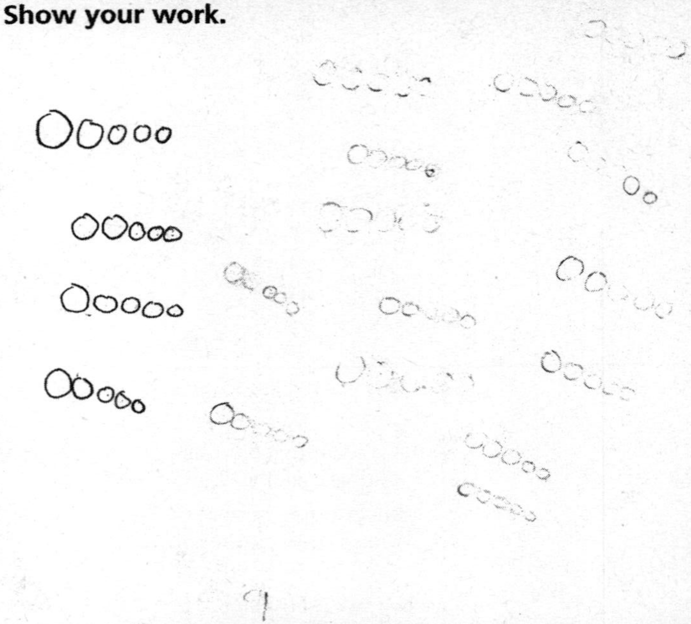

Level 1 The work shown is incomplete. Neither multiplication nor addition sentences are included.